MATH WORKBOOK

ADDITION

1) $131 + 103$ = ...

2) $249 + 221$ = ...

3) $245 + 220$ = ...

4) $216 + 215$ = ...

5) $185 + 108$ = ...

6) $263 + 223$ = ...

7) $126 + 156$ = ...

8) $275 + 244$ = ...

9) $192 + 113$ = ...

10) $158 + 196$ = ...

11) $118 + 244$ = ...

12) $158 + 176$ = ...

13) $173 + 262$ = ...

14) $293 + 162$ = ...

15) $203 + 202$ = ...

16) $197 + 258$ = ...

17) $104 + 257$ = ...

18) $145 + 104$ = ...

19) $236 + 263$ = ...

20) $173 + 152$ = ...

1) $228 + 155 =$ ....

2) $112 + 284 =$ ....

3) $211 + 104 =$ ....

4) $220 + 169 =$ ....

5) $110 + 271 =$ ....

6) $109 + 229 =$ ....

7) $240 + 107 =$ ....

8) $256 + 132 =$ ....

9) $128 + 213 =$ ....

10) $221 + 236 =$ ....

11) $283 + 235 =$ ....

12) $208 + 289 =$ ....

13) $243 + 145 =$ ....

14) $178 + 265 =$ ....

15) $280 + 259 =$ ....

16) $159 + 163 =$ ....

17) $215 + 189 =$ ....

18) $224 + 186 =$ ....

19) $273 + 233 =$ ....

20) $133 + 250 =$ ....

1) $226 + 255 =$ ....

2) $188 + 152 =$ ....

3) $212 + 246 =$ ....

4) $114 + 151 =$ ....

5) $183 + 170 =$ ....

6) $159 + 195 =$ ....

7) $180 + 271 =$ ....

8) $219 + 173 =$ ....

9) $187 + 226 =$ ....

10) $183 + 213 =$ ....

11) $234 + 104 =$ ....

12) $183 + 136 =$ ....

13) $199 + 127 =$ ....

14) $211 + 102 =$ ....

15) $189 + 231 =$ ....

16) $239 + 250 =$ ....

17) $191 + 267 =$ ....

18) $124 + 163 =$ ....

19) $121 + 296 =$ ....

20) $170 + 259 =$ ....

1) $347 + 220 =$ ....

2) $359 + 347 =$ ....

3) $274 + 113 =$ ....

4) $447 + 107 =$ ....

5) $379 + 304 =$ ....

6) $304 + 383 =$ ....

7) $412 + 297 =$ ....

8) $478 + 270 =$ ....

9) $362 + 101 =$ ....

10) $337 + 310 =$ ....

11) $404 + 428 =$ ....

12) $400 + 144 =$ ....

13) $268 + 273 =$ ....

14) $296 + 401 =$ ....

15) $345 + 299 =$ ....

16) $362 + 287 =$ ....

17) $298 + 154 =$ ....

18) $352 + 129 =$ ....

19) $479 + 316 =$ ....

20) $403 + 298 =$ ....

1) $\begin{array}{r} 414 \\ +\ 128 \\ \hline \end{array}$ ....

2) $\begin{array}{r} 360 \\ +\ 139 \\ \hline \end{array}$ ....

3) $\begin{array}{r} 492 \\ +\ 451 \\ \hline \end{array}$ ....

4) $\begin{array}{r} 314 \\ +\ 328 \\ \hline \end{array}$ ....

5) $\begin{array}{r} 496 \\ +\ 442 \\ \hline \end{array}$ ....

6) $\begin{array}{r} 349 \\ +\ 499 \\ \hline \end{array}$ ....

7) $\begin{array}{r} 298 \\ +\ 484 \\ \hline \end{array}$ ....

8) $\begin{array}{r} 448 \\ +\ 113 \\ \hline \end{array}$ ....

9) $\begin{array}{r} 297 \\ +\ 437 \\ \hline \end{array}$ ....

10) $\begin{array}{r} 348 \\ +\ 219 \\ \hline \end{array}$ ....

11) $\begin{array}{r} 360 \\ +\ 423 \\ \hline \end{array}$ ....

12) $\begin{array}{r} 419 \\ +\ 266 \\ \hline \end{array}$ ....

13) $\begin{array}{r} 397 \\ +\ 244 \\ \hline \end{array}$ ....

14) $\begin{array}{r} 326 \\ +\ 498 \\ \hline \end{array}$ ....

15) $\begin{array}{r} 412 \\ +\ 435 \\ \hline \end{array}$ ....

16) $\begin{array}{r} 332 \\ +\ 304 \\ \hline \end{array}$ ....

17) $\begin{array}{r} 246 \\ +\ 445 \\ \hline \end{array}$ ....

18) $\begin{array}{r} 435 \\ +\ 351 \\ \hline \end{array}$ ....

19) $\begin{array}{r} 234 \\ +\ 260 \\ \hline \end{array}$ ....

20) $\begin{array}{r} 456 \\ +\ 395 \\ \hline \end{array}$ ....

1) 337 + 254 = ...	2) 458 + 159 = ...	3) 324 + 494 = ...	4) 454 + 286 = ...
5) 254 + 180 = ...	6) 476 + 469 = ...	7) 497 + 403 = ...	8) 424 + 181 = ...
9) 284 + 325 = ...	10) 452 + 424 = ...	11) 353 + 233 = ...	12) 348 + 225 = ...
13) 347 + 360 = ...	14) 376 + 124 = ...	15) 374 + 450 = ...	16) 286 + 499 = ...
17) 488 + 196 = ...	18) 284 + 166 = ...	19) 341 + 251 = ...	20) 426 + 336 = ...

1) $320 + 414$ = ...

2) $465 + 222$ = ...

3) $474 + 469$ = ...

4) $489 + 400$ = ...

5) $344 + 333$ = ...

6) $420 + 461$ = ...

7) $442 + 200$ = ...

8) $443 + 149$ = ...

9) $327 + 176$ = ...

10) $451 + 126$ = ...

11) $455 + 267$ = ...

12) $415 + 301$ = ...

13) $275 + 303$ = ...

14) $415 + 231$ = ...

15) $242 + 162$ = ...

16) $282 + 468$ = ...

17) $416 + 495$ = ...

18) $257 + 186$ = ...

19) $351 + 342$ = ...

20) $317 + 322$ = ...

1) 537 + 969	2) 859 + 778	3) 310 + 961	4) 682 + 878 ....
5) 618 + 827	6) 893 + 326	7) 530 + 965	8) 331 + 542 ....
9) 675 + 396	10) 971 + 960	11) 543 + 464	12) 777 + 944 ....
13) 338 + 481	14) 505 + 404	15) 485 + 943	16) 651 + 866 ....
17) 547 + 826	18) 870 + 898	19) 567 + 570	20) 798 + 504 ....

1) 756 + 578 = ...	2) 693 + 674 = ...	3) 477 + 327 = ...	4) 777 + 703 = ...
5) 394 + 432 = ...	6) 973 + 406 = ...	7) 597 + 767 = ...	8) 894 + 964 = ...
9) 931 + 791 = ...	10) 537 + 421 = ...	11) 479 + 481 = ...	12) 997 + 952 = ...
13) 317 + 873 = ...	14) 897 + 516 = ...	15) 819 + 797 = ...	16) 707 + 653 = ...
17) 829 + 968 = ...	18) 747 + 786 = ...	19) 885 + 857 = ...	20) 809 + 680 = ...

1) 429 + 562 = ...

2) 883 + 596 = ...

3) 724 + 621 = ...

4) 576 + 628 = ...

5) 319 + 408 = ...

6) 379 + 374 = ...

7) 980 + 856 = ...

8) 981 + 880 = ...

9) 552 + 324 = ...

10) 693 + 497 = ...

11) 693 + 695 = ...

12) 380 + 777 = ...

13) 656 + 307 = ...

14) 975 + 510 = ...

15) 619 + 473 = ...

16) 506 + 445 = ...

17) 500 + 506 = ...

18) 551 + 876 = ...

19) 714 + 632 = ...

20) 549 + 937 = ...

1) 1,586 + 754 = ...

2) 1,310 + 816 = ...

3) 1,877 + 328 = ...

4) 1,051 + 408 = ...

5) 1,254 + 457 = ...

6) 1,493 + 906 = ...

7) 1,690 + 538 = ...

8) 1,122 + 632 = ...

9) 1,563 + 630 = ...

10) 1,284 + 649 = ...

11) 1,778 + 640 = ...

12) 1,088 + 421 = ...

13) 1,116 + 354 = ...

14) 1,973 + 437 = ...

15) 1,858 + 775 = ...

1) 1,067 + 1,976 = ...

2) 1,045 + 1,570 = ...

3) 1,284 + 1,612 = ...

4) 1,341 + 1,705 = ...

5) 1,121 + 1,467 = ...

6) 1,069 + 1,142 = ...

7) 1,189 + 1,421 = ...

8) 1,604 + 1,212 = ...

9) 1,398 + 1,699 = ...

10) 1,057 + 1,284 = ...

11) 1,102 + 1,584 = ...

12) 1,252 + 1,920 = ...

13) 1,271 + 1,129 = ...

14) 1,201 + 1,460 = ...

15) 1,008 + 1,569 = ...

1) 1,488 + 1,078 = ...

2) 1,856 + 1,227 = ...

3) 1,855 + 1,639 = ...

4) 1,029 + 1,550 = ...

5) 1,441 + 1,390 = ...

6) 1,482 + 1,724 = ...

7) 1,125 + 1,299 = ...

8) 1,155 + 1,411 = ...

9) 1,881 + 1,697 = ...

10) 1,552 + 1,416 = ...

11) 1,870 + 1,157 = ...

12) 1,534 + 1,710 = ...

13) 1,491 + 1,179 = ...

14) 1,143 + 1,091 = ...

15) 1,576 + 1,208 = ...

1) 1,389 + 1,908 = ...

2) 1,955 + 1,453 = ...

3) 1,069 + 1,195 = ...

4) 1,478 + 1,699 = ...

5) 1,866 + 1,874 = ...

6) 1,332 + 1,544 = ...

7) 1,146 + 1,737 = ...

8) 1,676 + 1,495 = ...

9) 1,938 + 1,973 = ...

10) 1,233 + 1,463 = ...

11) 1,149 + 1,754 = ...

12) 1,949 + 1,516 = ...

13) 1,104 + 1,067 = ...

14) 1,648 + 1,545 = ...

15) 1,621 + 1,395 = ...

1) 1,508 + 1,198 = ...

2) 1,496 + 1,476 = ...

3) 1,435 + 1,083 = ...

4) 1,675 + 1,312 = ...

5) 1,801 + 1,079 = ...

6) 1,635 + 1,959 = ...

7) 1,439 + 1,349 = ...

8) 1,003 + 1,547 = ...

9) 1,933 + 1,077 = ...

10) 1,472 + 1,653 = ...

11) 1,720 + 1,274 = ...

12) 1,665 + 1,635 = ...

13) 1,406 + 1,307 = ...

14) 1,301 + 1,466 = ...

15) 1,220 + 1,014 = ...

1) 1,311 + 1,749 = ...

2) 1,288 + 1,505 = ...

3) 1,467 + 1,516 = ...

4) 1,557 + 1,631 = ...

5) 1,648 + 1,731 = ...

6) 1,234 + 1,533 = ...

7) 1,861 + 1,071 = ...

8) 1,128 + 1,782 = ...

9) 1,209 + 1,092 = ...

10) 1,787 + 1,564 = ...

11) 1,983 + 1,312 = ...

12) 1,374 + 1,753 = ...

13) 1,134 + 1,394 = ...

14) 1,865 + 1,891 = ...

15) 1,160 + 1,108 = ...

1) 1,364 + 1,479 = ...

2) 2,183 + 1,681 = ...

3) 2,695 + 1,292 = ...

4) 2,098 + 1,459 = ...

5) 1,043 + 1,935 = ...

6) 1,128 + 1,599 = ...

7) 1,044 + 1,487 = ...

8) 2,522 + 1,089 = ...

9) 2,739 + 1,392 = ...

10) 1,464 + 1,520 = ...

11) 2,374 + 1,335 = ...

12) 1,288 + 1,571 = ...

13) 2,939 + 1,783 = ...

14) 2,665 + 1,840 = ...

15) 2,538 + 1,101 = ...

1) $\begin{array}{r} 2,006 \\ +\ 1,846 \\ \hline \end{array}$

....

2) $\begin{array}{r} 1,692 \\ +\ 1,869 \\ \hline \end{array}$

....

3) $\begin{array}{r} 2,364 \\ +\ 1,342 \\ \hline \end{array}$

....

4) $\begin{array}{r} 2,465 \\ +\ 1,192 \\ \hline \end{array}$

....

5) $\begin{array}{r} 1,329 \\ +\ 1,408 \\ \hline \end{array}$

....

6) $\begin{array}{r} 2,694 \\ +\ 1,334 \\ \hline \end{array}$

....

7) $\begin{array}{r} 1,041 \\ +\ 1,711 \\ \hline \end{array}$

....

8) $\begin{array}{r} 1,128 \\ +\ 1,459 \\ \hline \end{array}$

....

9) $\begin{array}{r} 2,417 \\ +\ 1,690 \\ \hline \end{array}$

....

10) $\begin{array}{r} 2,841 \\ +\ 1,631 \\ \hline \end{array}$

....

11) $\begin{array}{r} 1,432 \\ +\ 1,592 \\ \hline \end{array}$

....

12) $\begin{array}{r} 1,283 \\ +\ 1,099 \\ \hline \end{array}$

....

13) $\begin{array}{r} 1,942 \\ +\ 1,434 \\ \hline \end{array}$

....

14) $\begin{array}{r} 1,423 \\ +\ 1,950 \\ \hline \end{array}$

....

15) $\begin{array}{r} 2,834 \\ +\ 1,105 \\ \hline \end{array}$

....

1) 2,911 + 1,636 = ...

2) 2,893 + 1,378 = ...

3) 2,710 + 1,706 = ...

4) 1,326 + 1,098 = ...

5) 2,649 + 1,519 = ...

6) 2,823 + 1,704 = ...

7) 1,495 + 1,398 = ...

8) 1,497 + 1,130 = ...

9) 2,565 + 1,275 = ...

10) 1,119 + 1,575 = ...

11) 2,061 + 1,646 = ...

12) 1,369 + 1,245 = ...

13) 1,253 + 1,973 = ...

14) 2,475 + 1,603 = ...

15) 1,323 + 1,061 = ...

1) 1,467 + 1,239 = ...

2) 2,145 + 1,839 = ...

3) 2,259 + 1,788 = ...

4) 1,762 + 1,784 = ...

5) 2,794 + 1,011 = ...

6) 1,619 + 1,451 = ...

7) 2,403 + 1,251 = ...

8) 2,765 + 1,969 = ...

9) 2,121 + 1,936 = ...

10) 2,528 + 1,959 = ...

11) 1,325 + 1,034 = ...

12) 2,381 + 1,883 = ...

13) 1,901 + 1,865 = ...

14) 1,420 + 1,213 = ...

15) 2,322 + 1,904 = ...

SUBTRACTION

1) 201 - 170 = ...

2) 231 - 137 = ...

3) 358 - 153 = ...

4) 418 - 161 = ...

5) 443 - 110 = ...

6) 440 - 150 = ...

7) 331 - 177 = ...

8) 427 - 159 = ...

9) 221 - 194 = ...

10) 201 - 109 = ...

11) 381 - 166 = ...

12) 310 - 135 = ...

13) 373 - 171 = ...

14) 493 - 110 = ...

15) 324 - 196 = ...

16) 299 - 178 = ...

17) 455 - 161 = ...

18) 248 - 123 = ...

19) 288 - 107 = ...

20) 369 - 177 = ...

1) 367 - 163	2) 260 - 185	3) 432 - 102	4) 394 - 183 ....
5) 383 - 153	6) 290 - 200	7) 312 - 124	8) 273 - 196 ....
9) 257 - 172	10) 465 - 174	11) 301 - 149	12) 380 - 116 ....
13) 310 - 148	14) 349 - 177	15) 313 - 182	16) 469 - 154 ....
17) 283 - 174	18) 263 - 121	19) 334 - 162	20) 460 - 145 ....

1) $356 - 187 =$ ....

2) $336 - 179 =$ ....

3) $389 - 176 =$ ....

4) $415 - 163 =$ ....

5) $207 - 103 =$ ....

6) $353 - 157 =$ ....

7) $457 - 143 =$ ....

8) $365 - 180 =$ ....

9) $385 - 163 =$ ....

10) $260 - 177 =$ ....

11) $201 - 145 =$ ....

12) $482 - 110 =$ ....

13) $398 - 128 =$ ....

14) $402 - 122 =$ ....

15) $279 - 115 =$ ....

16) $256 - 140 =$ ....

17) $242 - 157 =$ ....

18) $480 - 132 =$ ....

19) $226 - 108 =$ ....

20) $303 - 120 =$ ....

1) 225 - 150	2) 309 - 188	3) 453 - 111	4) 329 - 129 ....
5) 348 - 123	6) 413 - 153	7) 285 - 200	8) 465 - 116 ....
9) 326 - 114	10) 303 - 107	11) 359 - 123	12) 415 - 158 ....
13) 409 - 178	14) 367 - 142	15) 285 - 186	16) 315 - 174 ....
17) 485 - 200	18) 235 - 139	19) 345 - 115	20) 487 - 165 ....

1) 366 - 101 = ...

2) 201 - 164 = ...

3) 456 - 197 = ...

4) 376 - 129 = ...

5) 464 - 150 = ...

6) 401 - 127 = ...

7) 254 - 169 = ...

8) 475 - 106 = ...

9) 286 - 144 = ...

10) 382 - 178 = ...

11) 399 - 171 = ...

12) 377 - 115 = ...

13) 260 - 159 = ...

14) 421 - 189 = ...

15) 253 - 126 = ...

16) 381 - 170 = ...

17) 490 - 190 = ...

18) 230 - 115 = ...

19) 340 - 131 = ...

20) 325 - 101 = ...

1) $270 - 170 =$ ....

2) $233 - 162 =$ ....

3) $468 - 197 =$ ....

4) $291 - 128 =$ ....

5) $376 - 177 =$ ....

6) $377 - 198 =$ ....

7) $208 - 160 =$ ....

8) $336 - 183 =$ ....

9) $422 - 125 =$ ....

10) $406 - 181 =$ ....

11) $491 - 142 =$ ....

12) $424 - 142 =$ ....

13) $410 - 127 =$ ....

14) $356 - 156 =$ ....

15) $349 - 125 =$ ....

16) $419 - 161 =$ ....

17) $461 - 115 =$ ....

18) $432 - 163 =$ ....

19) $416 - 135 =$ ....

20) $332 - 149 =$ ....

1) $335 - 174 =$ ....

2) $436 - 133 =$ ....

3) $210 - 198 =$ ....

4) $232 - 188 =$ ....

5) $448 - 182 =$ ....

6) $233 - 139 =$ ....

7) $444 - 119 =$ ....

8) $419 - 164 =$ ....

9) $405 - 137 =$ ....

10) $275 - 111 =$ ....

11) $488 - 194 =$ ....

12) $385 - 193 =$ ....

13) $244 - 157 =$ ....

14) $334 - 125 =$ ....

15) $456 - 195 =$ ....

16) $404 - 103 =$ ....

17) $330 - 182 =$ ....

18) $212 - 134 =$ ....

19) $364 - 140 =$ ....

20) $303 - 119 =$ ....

1) 265 - 202	2) 619 - 337	3) 914 - 263	4) 591 - 266
5) 382 - 161	6) 482 - 240	7) 611 - 231	8) 359 - 135
9) 368 - 180	10) 987 - 163	11) 561 - 112	12) 481 - 169
13) 366 - 224	14) 482 - 325	15) 561 - 274	16) 663 - 301
17) 822 - 326	18) 277 - 128	19) 906 - 160	20) 412 - 245

1) 351 - 286 = ...

2) 281 - 217 = ...

3) 708 - 305 = ...

4) 749 - 328 = ...

5) 471 - 113 = ...

6) 401 - 104 = ...

7) 227 - 164 = ...

8) 758 - 362 = ...

9) 995 - 240 = ...

10) 353 - 170 = ...

11) 461 - 271 = ...

12) 898 - 127 = ...

13) 594 - 105 = ...

14) 852 - 273 = ...

15) 697 - 117 = ...

16) 909 - 277 = ...

17) 692 - 351 = ...

18) 435 - 332 = ...

19) 582 - 220 = ...

20) 399 - 206 = ...

1)	893 - 133	2)	312 - 160	3)	325 - 249	4)	517 - 159
5)	366 - 346	6)	515 - 279	7)	373 - 328	8)	223 - 217
9)	299 - 108	10)	983 - 225	11)	587 - 240	12)	258 - 157
13)	209 - 109	14)	590 - 192	15)	786 - 302	16)	923 - 240
17)	394 - 285	18)	590 - 141	19)	390 - 147	20)	740 - 234

1) 1,240 - 104 = ...

2) 1,212 - 157 = ...

3) 1,443 - 110 = ...

4) 1,847 - 192 = ...

5) 1,356 - 125 = ...

6) 1,263 - 191 = ...

7) 1,959 - 354 = ...

8) 1,506 - 180 = ...

9) 1,690 - 223 = ...

10) 1,371 - 232 = ...

11) 1,924 - 260 = ...

12) 1,973 - 215 = ...

13) 1,491 - 214 = ...

14) 1,176 - 320 = ...

15) 1,075 - 278 = ...

1) 1,126 - 1,123 = ...

2) 1,715 - 1,161 = ...

3) 1,979 - 1,095 = ...

4) 1,234 - 1,026 = ...

5) 1,156 - 1,059 = ...

6) 1,773 - 1,110 = ...

7) 1,925 - 1,394 = ...

8) 1,753 - 1,434 = ...

9) 1,781 - 1,230 = ...

10) 1,522 - 1,358 = ...

11) 1,181 - 1,025 = ...

12) 1,829 - 1,191 = ...

13) 1,226 - 1,153 = ...

14) 1,949 - 1,219 = ...

15) 1,897 - 1,472 = ...

1) 1,614 - 1,490 = ...

2) 1,352 - 1,187 = ...

3) 1,505 - 1,322 = ...

4) 1,985 - 1,276 = ...

5) 1,080 - 1,024 = ...

6) 1,988 - 1,074 = ...

7) 1,333 - 1,032 = ...

8) 1,301 - 1,073 = ...

9) 1,797 - 1,108 = ...

10) 1,213 - 1,047 = ...

11) 1,593 - 1,216 = ...

12) 1,787 - 1,299 = ...

13) 1,867 - 1,487 = ...

14) 1,250 - 1,059 = ...

15) 1,521 - 1,108 = ...

1) 1,738 - 1,280 = ...

2) 1,435 - 1,401 = ...

3) 1,760 - 1,351 = ...

4) 1,093 - 1,011 = ...

5) 1,591 - 1,021 = ...

6) 1,181 - 1,035 = ...

7) 1,753 - 1,332 = ...

8) 1,530 - 1,196 = ...

9) 1,748 - 1,015 = ...

10) 1,864 - 1,213 = ...

11) 1,180 - 1,024 = ...

12) 1,587 - 1,242 = ...

13) 1,604 - 1,322 = ...

14) 1,891 - 1,412 = ...

15) 1,913 - 1,026 = ...

1) $\begin{array}{r} 1,505 \\ -\ 1,422 \\ \hline \end{array}$

....

2) $\begin{array}{r} 1,242 \\ -\ 1,208 \\ \hline \end{array}$

....

3) $\begin{array}{r} 1,907 \\ -\ 1,211 \\ \hline \end{array}$

....

4) $\begin{array}{r} 1,598 \\ -\ 1,421 \\ \hline \end{array}$

....

5) $\begin{array}{r} 1,415 \\ -\ 1,009 \\ \hline \end{array}$

....

6) $\begin{array}{r} 1,607 \\ -\ 1,486 \\ \hline \end{array}$

....

7) $\begin{array}{r} 1,952 \\ -\ 1,159 \\ \hline \end{array}$

....

8) $\begin{array}{r} 1,115 \\ -\ 1,093 \\ \hline \end{array}$

....

9) $\begin{array}{r} 1,268 \\ -\ 1,055 \\ \hline \end{array}$

....

10) $\begin{array}{r} 1,538 \\ -\ 1,039 \\ \hline \end{array}$

....

11) $\begin{array}{r} 1,455 \\ -\ 1,112 \\ \hline \end{array}$

....

12) $\begin{array}{r} 1,356 \\ -\ 1,191 \\ \hline \end{array}$

....

13) $\begin{array}{r} 1,115 \\ -\ 1,066 \\ \hline \end{array}$

....

14) $\begin{array}{r} 1,432 \\ -\ 1,116 \\ \hline \end{array}$

....

15) $\begin{array}{r} 1,350 \\ -\ 1,226 \\ \hline \end{array}$

....

1) 1,442 - 1,382 = ...

2) 1,998 - 1,323 = ...

3) 1,701 - 1,256 = ...

4) 1,488 - 1,433 = ...

5) 1,627 - 1,488 = ...

6) 1,259 - 1,126 = ...

7) 1,707 - 1,373 = ...

8) 1,329 - 1,325 = ...

9) 1,329 - 1,162 = ...

10) 1,770 - 1,266 = ...

11) 1,900 - 1,045 = ...

12) 1,899 - 1,110 = ...

13) 1,315 - 1,052 = ...

14) 1,305 - 1,060 = ...

15) 1,907 - 1,356 = ...

1) 1,313 - 1,295 = ...

2) 1,217 - 1,147 = ...

3) 1,642 - 1,019 = ...

4) 1,390 - 1,297 = ...

5) 1,844 - 1,311 = ...

6) 1,352 - 1,340 = ...

7) 1,779 - 1,166 = ...

8) 1,537 - 1,103 = ...

9) 1,290 - 1,284 = ...

10) 1,916 - 1,202 = ...

11) 1,544 - 1,254 = ...

12) 1,693 - 1,037 = ...

13) 1,614 - 1,277 = ...

14) 1,707 - 1,422 = ...

15) 1,635 - 1,265 = ...

1) 2,946 - 1,471 = ...

2) 1,913 - 1,088 = ...

3) 1,686 - 1,317 = ...

4) 2,652 - 1,411 = ...

5) 2,699 - 1,402 = ...

6) 2,213 - 1,128 = ...

7) 2,920 - 1,399 = ...

8) 2,617 - 1,470 = ...

9) 1,250 - 1,025 = ...

10) 1,902 - 1,352 = ...

11) 1,862 - 1,473 = ...

12) 1,785 - 1,099 = ...

13) 1,379 - 1,347 = ...

14) 2,634 - 1,210 = ...

15) 1,988 - 1,151 = ...

1) $$\begin{array}{r} 2,284 \\ -\ 1,228 \\ \hline \end{array}$$

....

2) $$\begin{array}{r} 2,866 \\ -\ 1,450 \\ \hline \end{array}$$

....

3) $$\begin{array}{r} 2,757 \\ -\ 1,058 \\ \hline \end{array}$$

....

4) $$\begin{array}{r} 1,792 \\ -\ 1,349 \\ \hline \end{array}$$

....

5) $$\begin{array}{r} 1,760 \\ -\ 1,383 \\ \hline \end{array}$$

....

6) $$\begin{array}{r} 1,052 \\ -\ 1,029 \\ \hline \end{array}$$

....

7) $$\begin{array}{r} 1,517 \\ -\ 1,494 \\ \hline \end{array}$$

....

8) $$\begin{array}{r} 1,857 \\ -\ 1,409 \\ \hline \end{array}$$

....

9) $$\begin{array}{r} 1,813 \\ -\ 1,164 \\ \hline \end{array}$$

....

10) $$\begin{array}{r} 2,097 \\ -\ 1,390 \\ \hline \end{array}$$

....

11) $$\begin{array}{r} 2,225 \\ -\ 1,470 \\ \hline \end{array}$$

....

12) $$\begin{array}{r} 2,654 \\ -\ 1,228 \\ \hline \end{array}$$

....

13) $$\begin{array}{r} 2,263 \\ -\ 1,314 \\ \hline \end{array}$$

....

14) $$\begin{array}{r} 2,173 \\ -\ 1,100 \\ \hline \end{array}$$

....

15) $$\begin{array}{r} 2,726 \\ -\ 1,469 \\ \hline \end{array}$$

....

1) 2,253 - 1,028 = ...

2) 2,102 - 1,235 = ...

3) 2,541 - 1,180 = ...

4) 1,448 - 1,405 = ...

5) 1,336 - 1,217 = ...

6) 2,402 - 1,315 = ...

7) 1,956 - 1,134 = ...

8) 1,503 - 1,265 = ...

9) 1,393 - 1,328 = ...

10) 1,884 - 1,443 = ...

11) 2,063 - 1,053 = ...

12) 1,637 - 1,220 = ...

13) 2,941 - 1,467 = ...

14) 2,772 - 1,356 = ...

15) 1,727 - 1,076 = ...

ANSWERS

Page 1

1) 131 + 103 = 234 | 2) 249 + 221 = 470 | 3) 245 + 220 = 465 | 4) 216 + 215 = 431

5) 185 + 108 = 293 | 6) 263 + 223 = 486 | 7) 126 + 156 = 282 | 8) 275 + 244 = 519

9) 192 + 113 = 305 | 10) 158 + 196 = 354 | 11) 118 + 244 = 362 | 12) 158 + 176 = 334

13) 173 + 262 = 435 | 14) 293 + 162 = 455 | 15) 203 + 202 = 405 | 16) 197 + 258 = 455

17) 104 + 257 = 361 | 18) 145 + 104 = 249 | 19) 236 + 263 = 499 | 20) 173 + 152 = 325

Page 2

1) 228 + 155 = 383 | 2) 112 + 284 = 396 | 3) 211 + 104 = 315 | 4) 220 + 169 = 389

5) 110 + 271 = 381 | 6) 109 + 229 = 338 | 7) 240 + 107 = 347 | 8) 256 + 132 = 388

9) 128 + 213 = 341 | 10) 221 + 236 = 457 | 11) 283 + 235 = 518 | 12) 208 + 289 = 497

13) 243 + 145 = 388 | 14) 178 + 265 = 443 | 15) 280 + 259 = 539 | 16) 159 + 163 = 322

17) 215 + 189 = 404 | 18) 224 + 186 = 410 | 19) 273 + 233 = 506 | 20) 133 + 250 = 383

Page 3

1) 226 + 255 = 481 | 2) 188 + 152 = 340 | 3) 212 + 246 = 458 | 4) 114 + 151 = 265

5) 183 + 170 = 353 | 6) 159 + 195 = 354 | 7) 180 + 271 = 451 | 8) 219 + 173 = 392

9) 187 + 226 = 413 | 10) 183 + 213 = 396 | 11) 234 + 104 = 338 | 12) 183 + 136 = 319

13) 199 + 127 = 326 | 14) 211 + 102 = 313 | 15) 189 + 231 = 420 | 16) 239 + 250 = 489

17) 191 + 267 = 458 | 18) 124 + 163 = 287 | 19) 121 + 296 = 417 | 20) 170 + 259 = 429

Page 4

1) 347 + 220 = 567 | 2) 359 + 347 = 706 | 3) 274 + 113 = 387 | 4) 447 + 107 = 554

5) 379 + 304 = 683 | 6) 304 + 383 = 687 | 7) 412 + 297 = 709 | 8) 478 + 270 = 748

9) 362 + 101 = 463 | 10) 337 + 310 = 647 | 11) 404 + 428 = 832 | 12) 400 + 144 = 544

13) 268 + 273 = 541 | 14) 296 + 401 = 697 | 15) 345 + 299 = 644 | 16) 362 + 287 = 649

17) 298 + 154 = 452 | 18) 352 + 129 = 481 | 19) 479 + 316 = 795 | 20) 403 + 298 = 701

Page 5

1) 414 + 128 542	2) 360 + 139 499	3) 492 + 451 943	4) 314 + 328 642
5) 496 + 442 938	6) 349 + 499 848	7) 298 + 484 782	8) 448 + 113 561
9) 297 + 437 734	10) 348 + 219 567	11) 360 + 423 783	12) 419 + 266 685
13) 397 + 244 641	14) 326 + 498 824	15) 412 + 435 847	16) 332 + 304 636
17) 246 + 445 691	18) 435 + 351 786	19) 234 + 260 494	20) 456 + 395 851

Page 6

1) 337 + 254 591	2) 458 + 159 617	3) 324 + 494 818	4) 454 + 286 740
5) 254 + 180 434	6) 476 + 469 945	7) 497 + 403 900	8) 424 + 181 605
9) 284 + 325 609	10) 452 + 424 876	11) 353 + 233 586	12) 348 + 225 573
13) 347 + 360 707	14) 376 + 124 500	15) 374 + 450 824	16) 286 + 499 785
17) 488 + 196 684	18) 284 + 166 450	19) 341 + 251 592	20) 426 + 336 762

Page 7

1) 320 + 414 734	2) 465 + 222 687	3) 474 + 469 943	4) 489 + 400 889
5) 344 + 333 677	6) 420 + 461 881	7) 442 + 200 642	8) 443 + 149 592
9) 327 + 176 503	10) 451 + 126 577	11) 455 + 267 722	12) 415 + 301 716
13) 275 + 303 578	14) 415 + 231 646	15) 242 + 162 404	16) 282 + 468 750
17) 416 + 495 911	18) 257 + 186 443	19) 351 + 342 693	20) 317 + 322 639

Page 8

1) 537 + 969 1,506	2) 859 + 778 1,637	3) 310 + 961 1,271	4) 682 + 878 1,560
5) 618 + 827 1,445	6) 893 + 326 1,219	7) 530 + 965 1,495	8) 331 + 542 873
9) 675 + 396 1,071	10) 971 + 960 1,931	11) 543 + 464 1,007	12) 777 + 944 1,721
13) 338 + 481 819	14) 505 + 404 909	15) 485 + 943 1,428	16) 651 + 866 1,517
17) 547 + 826 1,373	18) 870 + 898 1,768	19) 567 + 570 1,137	20) 798 + 504 1,302

Page 9

1) 756 + 578 1,334	2) 693 + 674 1,367	3) 477 + 327 804	4) 777 + 703 1,480
5) 394 + 432 826	6) 973 + 406 1,379	7) 597 + 767 1,364	8) 894 + 964 1,858
9) 931 + 791 1,722	10) 537 + 421 958	11) 479 + 481 960	12) 997 + 952 1,949
13) 317 + 873 1,190	14) 897 + 516 1,413	15) 819 + 797 1,616	16) 707 + 653 1,360
17) 829 + 968 1,797	18) 747 + 786 1,533	19) 885 + 857 1,742	20) 809 + 680 1,489

Page 10

1) 429 + 562 991	2) 883 + 596 1,479	3) 724 + 621 1,345	4) 576 + 628 1,204
5) 319 + 408 727	6) 379 + 374 753	7) 980 + 856 1,836	8) 981 + 880 1,861
9) 552 + 324 876	10) 693 + 497 1,190	11) 693 + 695 1,388	12) 380 + 777 1,157
13) 656 + 307 963	14) 975 + 510 1,485	15) 619 + 473 1,092	16) 506 + 445 951
17) 500 + 506 1,006	18) 551 + 876 1,427	19) 714 + 632 1,346	20) 549 + 937 1,486

Page 11

1) 1,586 + 754 2,340	2) 1,310 + 816 2,126	3) 1,877 + 328 2,205
4) 1,051 + 408 1,459	5) 1,254 + 457 1,711	6) 1,493 + 906 2,399
7) 1,690 + 538 2,228	8) 1,122 + 632 1,754	9) 1,563 + 630 2,193
10) 1,284 + 649 1,933	11) 1,778 + 640 2,418	12) 1,088 + 421 1,509
13) 1,116 + 354 1,470	14) 1,973 + 437 2,410	15) 1,858 + 775 2,633

Page 12

1) 1,067 + 1,976 3,043	2) 1,045 + 1,570 2,615	3) 1,284 + 1,612 2,896
4) 1,341 + 1,705 3,046	5) 1,121 + 1,467 2,588	6) 1,069 + 1,142 2,211
7) 1,189 + 1,421 2,610	8) 1,604 + 1,212 2,816	9) 1,398 + 1,699 3,097
10) 1,057 + 1,284 2,341	11) 1,102 + 1,584 2,686	12) 1,252 + 1,920 3,172
13) 1,271 + 1,129 2,400	14) 1,201 + 1,460 2,661	15) 1,008 + 1,569 2,577

Page 13

1) 1,488 + 1,078 2,566
2) 1,856 + 1,227 3,083
3) 1,855 + 1,639 3,494
4) 1,029 + 1,550 2,579
5) 1,441 + 1,390 2,831
6) 1,482 + 1,724 3,206
7) 1,125 + 1,299 2,424
8) 1,155 + 1,411 2,566
9) 1,881 + 1,697 3,578
10) 1,552 + 1,416 2,968
11) 1,870 + 1,157 3,027
12) 1,534 + 1,710 3,244
13) 1,491 + 1,179 2,670
14) 1,143 + 1,091 2,234
15) 1,576 + 1,208 2,784

Page 14

1) 1,389 + 1,908 3,297
2) 1,955 + 1,453 3,408
3) 1,069 + 1,195 2,264
4) 1,478 + 1,699 3,177
5) 1,866 + 1,874 3,740
6) 1,332 + 1,544 2,876
7) 1,146 + 1,737 2,883
8) 1,676 + 1,495 3,171
9) 1,938 + 1,973 3,911
10) 1,233 + 1,463 2,696
11) 1,149 + 1,754 2,903
12) 1,949 + 1,516 3,465
13) 1,104 + 1,067 2,171
14) 1,648 + 1,545 3,193
15) 1,621 + 1,395 3,016

Page 15

1) 1,508 + 1,198 2,706
2) 1,496 + 1,476 2,972
3) 1,435 + 1,083 2,518
4) 1,675 + 1,312 2,987
5) 1,801 + 1,079 2,880
6) 1,635 + 1,959 3,594
7) 1,439 + 1,349 2,788
8) 1,003 + 1,547 2,550
9) 1,933 + 1,077 3,010
10) 1,472 + 1,653 3,125
11) 1,720 + 1,274 2,994
12) 1,665 + 1,635 3,300
13) 1,406 + 1,307 2,713
14) 1,301 + 1,466 2,767
15) 1,220 + 1,014 2,234

Page 16

1) 1,311 + 1,749 3,060
2) 1,288 + 1,505 2,793
3) 1,467 + 1,516 2,983
4) 1,557 + 1,631 3,188
5) 1,648 + 1,731 3,379
6) 1,234 + 1,533 2,767
7) 1,861 + 1,071 2,932
8) 1,128 + 1,782 2,910
9) 1,209 + 1,092 2,301
10) 1,787 + 1,564 3,351
11) 1,983 + 1,312 3,295
12) 1,374 + 1,753 3,127
13) 1,134 + 1,394 2,528
14) 1,865 + 1,891 3,756
15) 1,160 + 1,108 2,268

Page 17

1) 1,364 + 1,479 = 2,843
2) 2,183 + 1,681 = 3,864
3) 2,695 + 1,292 = 3,987
4) 2,098 + 1,459 = 3,557
5) 1,043 + 1,935 = 2,978
6) 1,128 + 1,599 = 2,727
7) 1,044 + 1,487 = 2,531
8) 2,522 + 1,089 = 3,611
9) 2,739 + 1,392 = 4,131
10) 1,464 + 1,520 = 2,984
11) 2,374 + 1,335 = 3,709
12) 1,288 + 1,571 = 2,859
13) 2,939 + 1,783 = 4,722
14) 2,665 + 1,840 = 4,505
15) 2,538 + 1,101 = 3,639

Page 18

1) 2,006 + 1,846 = 3,852
2) 1,692 + 1,869 = 3,561
3) 2,364 + 1,342 = 3,706
4) 2,465 + 1,192 = 3,657
5) 1,329 + 1,408 = 2,737
6) 2,694 + 1,334 = 4,028
7) 1,041 + 1,711 = 2,752
8) 1,128 + 1,459 = 2,587
9) 2,417 + 1,690 = 4,107
10) 2,841 + 1,631 = 4,472
11) 1,432 + 1,592 = 3,024
12) 1,283 + 1,099 = 2,382
13) 1,942 + 1,434 = 3,376
14) 1,423 + 1,950 = 3,373
15) 2,834 + 1,105 = 3,939

Page 19

1) 2,911 + 1,636 = 4,547
2) 2,893 + 1,378 = 4,271
3) 2,710 + 1,706 = 4,416
4) 1,326 + 1,098 = 2,424
5) 2,649 + 1,519 = 4,168
6) 2,823 + 1,704 = 4,527
7) 1,495 + 1,398 = 2,893
8) 1,497 + 1,130 = 2,627
9) 2,565 + 1,275 = 3,840
10) 1,119 + 1,575 = 2,694
11) 2,061 + 1,646 = 3,707
12) 1,369 + 1,245 = 2,614
13) 1,253 + 1,973 = 3,226
14) 2,475 + 1,603 = 4,078
15) 1,323 + 1,061 = 2,384

Page 20

1) 1,467 + 1,239 = 2,706
2) 2,145 + 1,839 = 3,984
3) 2,259 + 1,788 = 4,047
4) 1,762 + 1,784 = 3,546
5) 2,794 + 1,011 = 3,805
6) 1,619 + 1,451 = 3,070
7) 2,403 + 1,251 = 3,654
8) 2,765 + 1,969 = 4,734
9) 2,121 + 1,936 = 4,057
10) 2,528 + 1,959 = 4,487
11) 1,325 + 1,034 = 2,359
12) 2,381 + 1,883 = 4,264
13) 1,901 + 1,865 = 3,766
14) 1,420 + 1,213 = 2,633
15) 2,322 + 1,904 = 4,226

Page 21

1) 201 - 170 = 31
2) 231 - 137 = 94
3) 358 - 153 = 205
4) 418 - 161 = 257
5) 443 - 110 = 333
6) 440 - 150 = 290
7) 331 - 177 = 154
8) 427 - 159 = 268
9) 221 - 194 = 27
10) 201 - 109 = 92
11) 381 - 166 = 215
12) 310 - 135 = 175
13) 373 - 171 = 202
14) 493 - 110 = 383
15) 324 - 196 = 128
16) 299 - 178 = 121
17) 455 - 161 = 294
18) 248 - 123 = 125
19) 288 - 107 = 181
20) 369 - 177 = 192

Page 22

1) 367 - 163 = 204
2) 260 - 185 = 75
3) 432 - 102 = 330
4) 394 - 183 = 211
5) 383 - 153 = 230
6) 290 - 200 = 90
7) 312 - 124 = 188
8) 273 - 196 = 77
9) 257 - 172 = 85
10) 465 - 174 = 291
11) 301 - 149 = 152
12) 380 - 116 = 264
13) 310 - 148 = 162
14) 349 - 177 = 172
15) 313 - 182 = 131
16) 469 - 154 = 315
17) 283 - 174 = 109
18) 263 - 121 = 142
19) 334 - 162 = 172
20) 460 - 145 = 315

Page 23

1) 356 - 187 = 169
2) 336 - 179 = 157
3) 389 - 176 = 213
4) 415 - 163 = 252
5) 207 - 103 = 104
6) 353 - 157 = 196
7) 457 - 143 = 314
8) 365 - 180 = 185
9) 385 - 163 = 222
10) 260 - 177 = 83
11) 201 - 145 = 56
12) 482 - 110 = 372
13) 398 - 128 = 270
14) 402 - 122 = 280
15) 279 - 115 = 164
16) 256 - 140 = 116
17) 242 - 157 = 85
18) 480 - 132 = 348
19) 226 - 108 = 118
20) 303 - 120 = 183

Page 24

1) 225 - 150 = 75
2) 309 - 188 = 121
3) 453 - 111 = 342
4) 329 - 129 = 200
5) 348 - 123 = 225
6) 413 - 153 = 260
7) 285 - 200 = 85
8) 465 - 116 = 349
9) 326 - 114 = 212
10) 303 - 107 = 196
11) 359 - 123 = 236
12) 415 - 158 = 257
13) 409 - 178 = 231
14) 367 - 142 = 225
15) 285 - 186 = 99
16) 315 - 174 = 141
17) 485 - 200 = 285
18) 235 - 139 = 96
19) 345 - 115 = 230
20) 487 - 165 = 322

Page 25

1) 366 - 101 265	2) 201 - 164 37	3) 456 - 197 259	4) 376 - 129 247
5) 464 - 150 314	6) 401 - 127 274	7) 254 - 169 85	8) 475 - 106 369
9) 286 - 144 142	10) 382 - 178 204	11) 399 - 171 228	12) 377 - 115 262
13) 260 - 159 101	14) 421 - 189 232	15) 253 - 126 127	16) 381 - 170 211
17) 490 - 190 300	18) 230 - 115 115	19) 340 - 131 209	20) 325 - 101 224

Page 26

1) 270 - 170 100	2) 233 - 162 71	3) 468 - 197 271	4) 291 - 128 163
5) 376 - 177 199	6) 377 - 198 179	7) 208 - 160 48	8) 336 - 183 153
9) 422 - 125 297	10) 406 - 181 225	11) 491 - 142 349	12) 424 - 142 282
13) 410 - 127 283	14) 356 - 156 200	15) 349 - 125 224	16) 419 - 161 258
17) 461 - 115 346	18) 432 - 163 269	19) 416 - 135 281	20) 332 - 149 183

Page 27

1) 335 - 174 161	2) 436 - 133 303	3) 210 - 198 12	4) 232 - 188 44
5) 448 - 182 266	6) 233 - 139 94	7) 444 - 119 325	8) 419 - 164 255
9) 405 - 137 268	10) 275 - 111 164	11) 488 - 194 294	12) 385 - 193 192
13) 244 - 157 87	14) 334 - 125 209	15) 456 - 195 261	16) 404 - 103 301
17) 330 - 182 148	18) 212 - 134 78	19) 364 - 140 224	20) 303 - 119 184

Page 28

1) 265 - 202 63	2) 619 - 337 282	3) 914 - 263 651	4) 591 - 266 325
5) 382 - 161 221	6) 482 - 240 242	7) 611 - 231 380	8) 359 - 135 224
9) 368 - 180 188	10) 987 - 163 824	11) 561 - 112 449	12) 481 - 169 312
13) 366 - 224 142	14) 482 - 325 157	15) 561 - 274 287	16) 663 - 301 362
17) 822 - 326 496	18) 277 - 128 149	19) 906 - 160 746	20) 412 - 245 167

Page 29

1) 351 - 286 65	2) 281 - 217 64	3) 708 - 305 403	4) 749 - 328 421
5) 471 - 113 358	6) 401 - 104 297	7) 227 - 164 63	8) 758 - 362 396
9) 995 - 240 755	10) 353 - 170 183	11) 461 - 271 190	12) 898 - 127 771
13) 594 - 105 489	14) 852 - 273 579	15) 697 - 117 580	16) 909 - 277 632
17) 692 - 351 341	18) 435 - 332 103	19) 582 - 220 362	20) 399 - 206 193

Page 30

1) 893 - 133 760	2) 312 - 160 152	3) 325 - 249 76	4) 517 - 159 358
5) 366 - 346 20	6) 515 - 279 236	7) 373 - 328 45	8) 223 - 217 6
9) 299 - 108 191	10) 983 - 225 758	11) 587 - 240 347	12) 258 - 157 101
13) 209 - 109 100	14) 590 - 192 398	15) 786 - 302 484	16) 923 - 240 683
17) 394 - 285 109	18) 590 - 141 449	19) 390 - 147 243	20) 740 - 234 506

Page 31

1) 1,240 - 104 1,136	2) 1,212 - 157 1,055	3) 1,443 - 110 1,333
4) 1,847 - 192 1,655	5) 1,356 - 125 1,231	6) 1,263 - 191 1,072
7) 1,959 - 354 1,605	8) 1,506 - 180 1,326	9) 1,690 - 223 1,467
10) 1,371 - 232 1,139	11) 1,924 - 260 1,664	12) 1,973 - 215 1,758
13) 1,491 - 214 1,277	14) 1,176 - 320 856	15) 1,075 - 278 797

Page 32

1) 1,126 - 1,123 3	2) 1,715 - 1,161 554	3) 1,979 - 1,095 884
4) 1,234 - 1,026 208	5) 1,156 - 1,059 97	6) 1,773 - 1,110 663
7) 1,925 - 1,394 531	8) 1,753 - 1,434 319	9) 1,781 - 1,230 551
10) 1,522 - 1,358 164	11) 1,181 - 1,025 156	12) 1,829 - 1,191 638
13) 1,226 - 1,153 73	14) 1,949 - 1,219 730	15) 1,897 - 1,472 425

Page 33

1)	1,614	- 1,490	124
2)	1,352	- 1,187	165
3)	1,505	- 1,322	183
4)	1,985	- 1,276	709
5)	1,080	- 1,024	56
6)	1,988	- 1,074	914
7)	1,333	- 1,032	301
8)	1,301	- 1,073	228
9)	1,797	- 1,108	689
10)	1,213	- 1,047	166
11)	1,593	- 1,216	377
12)	1,787	- 1,299	488
13)	1,867	- 1,487	380
14)	1,250	- 1,059	191
15)	1,521	- 1,108	413

Page 34

1)	1,738	- 1,280	458
2)	1,435	- 1,401	34
3)	1,760	- 1,351	409
4)	1,093	- 1,011	82
5)	1,591	- 1,021	570
6)	1,181	- 1,035	146
7)	1,753	- 1,332	421
8)	1,530	- 1,196	334
9)	1,748	- 1,015	733
10)	1,864	- 1,213	651
11)	1,180	- 1,024	156
12)	1,587	- 1,242	345
13)	1,604	- 1,322	282
14)	1,891	- 1,412	479
15)	1,913	- 1,026	887

Page 35

1)	1,505	- 1,422	83
2)	1,242	- 1,208	34
3)	1,907	- 1,211	696
4)	1,598	- 1,421	177
5)	1,415	- 1,009	406
6)	1,607	- 1,486	121
7)	1,952	- 1,159	793
8)	1,115	- 1,093	22
9)	1,268	- 1,055	213
10)	1,538	- 1,039	499
11)	1,455	- 1,112	343
12)	1,356	- 1,191	165
13)	1,115	- 1,066	49
14)	1,432	- 1,116	316
15)	1,350	- 1,226	124

Page 36

1)	1,442	- 1,382	60
2)	1,998	- 1,323	675
3)	1,701	- 1,256	445
4)	1,488	- 1,433	55
5)	1,627	- 1,488	139
6)	1,259	- 1,126	133
7)	1,707	- 1,373	334
8)	1,329	- 1,325	4
9)	1,329	- 1,162	167
10)	1,770	- 1,266	504
11)	1,900	- 1,045	855
12)	1,899	- 1,110	789
13)	1,315	- 1,052	263
14)	1,305	- 1,060	245
15)	1,907	- 1,356	551

Page 37

1)	1,313 - 1,295	18
2)	1,217 - 1,147	70
3)	1,642 - 1,019	623
4)	1,390 - 1,297	93
5)	1,844 - 1,311	533
6)	1,352 - 1,340	12
7)	1,779 - 1,166	613
8)	1,537 - 1,103	434
9)	1,290 - 1,284	6
10)	1,916 - 1,202	714
11)	1,544 - 1,254	290
12)	1,693 - 1,037	656
13)	1,614 - 1,277	337
14)	1,707 - 1,422	285
15)	1,635 - 1,265	370

Page 38

1)	2,946 - 1,471	1,475
2)	1,913 - 1,088	825
3)	1,686 - 1,317	369
4)	2,652 - 1,411	1,241
5)	2,699 - 1,402	1,297
6)	2,213 - 1,128	1,085
7)	2,920 - 1,399	1,521
8)	2,617 - 1,470	1,147
9)	1,250 - 1,025	225
10)	1,902 - 1,352	550
11)	1,862 - 1,473	389
12)	1,785 - 1,099	686
13)	1,379 - 1,347	32
14)	2,634 - 1,210	1,424
15)	1,988 - 1,151	837

Page 39

1)	2,284 - 1,228	1,056
2)	2,866 - 1,450	1,416
3)	2,757 - 1,058	1,699
4)	1,792 - 1,349	443
5)	1,760 - 1,383	377
6)	1,052 - 1,029	23
7)	1,517 - 1,494	23
8)	1,857 - 1,409	448
9)	1,813 - 1,164	649
10)	2,097 - 1,390	707
11)	2,225 - 1,470	755
12)	2,654 - 1,228	1,426
13)	2,263 - 1,314	949
14)	2,173 - 1,100	1,073
15)	2,726 - 1,469	1,257

Page 40

1)	2,253 - 1,028	1,225
2)	2,102 - 1,235	867
3)	2,541 - 1,180	1,361
4)	1,448 - 1,405	43
5)	1,336 - 1,217	119
6)	2,402 - 1,315	1,087
7)	1,956 - 1,134	822
8)	1,503 - 1,265	238
9)	1,393 - 1,328	65
10)	1,884 - 1,443	441
11)	2,063 - 1,053	1,010
12)	1,637 - 1,220	417
13)	2,941 - 1,467	1,474
14)	2,772 - 1,356	1,416
15)	1,727 - 1,076	651

Made in the USA
Coppell, TX
10 December 2024

37. Beauty and hygiene
38. House
39. Home
40. Furniture and interior
41. Materials
42. Traveling
43. Accommodation
44. Freetime and hobbies
45. Sports
46. Movies
47. Computer
48. Internet and social media
49. Photography
50. Music
51. Literature
52. Nightlife
53. Dating life
54. Law
55. Politics
56. Economy and finance
57. Religion and beliefs
58. Science
59. Space
60. Global warming
61. History
62. Pronouns
63. Pronouns 2
64. Top 30 verbs
65. Important verbs
66. Comparative
67. Comparing
68. Imperative
69. Conjunctions 1
70. Conjunctions 2
71. Adverbs
72. Structures

Kustantaja: BoD - Books on Demand, Helsinki, Suomi
Valmistaja: BoD - Books on Demand, Norderstedt, Saksa

ISBN: 9789528008989

1. Greetings

Welcome

Hello

Good morning

Good day

Good evening

Good night

Long time no see

Happy Birthday

How are you?

I'm fine

Ok

So so

Not good

Are you ok?

Yes

No

Thank you

No problem

You're welcome

Sorry

Excuse me

Nice to meet you

My pleasure

You too

Congratulations

Well done

Take care

Good luck

Goodbye

See you later

2. Introduction

What's your name?

My name is

You can call me ...

Who are you ?

Let me introduce myself

I am ...

How old are you?

I'm ... years old

Age

Year

First name

Nickname

Family name

Mr

Mrs

Miss

To be called

To call

To introduce

3. Numbers

One	First
Two	Second
Three	Third
Four	Fourth
Five	Fifth
Six	Sixth
Seven	Seventh
Eight	Eighth
Nine	Ninth
Ten	Tenth
Eleven	Eleventh
Twelve	Twelfth
Thirteen	Thirteenth
Fourteen	Fourteenth
Fifteen	Fifteenth
Sixteen	Sixteenth
Seventeen	Seventeeth
Eighteen	Eighteenth
Nineteen	Nineteenth
Twenty	Twentieth
To count	Equals to (is)
Plus	Minus
To divide	Divided by
To multiply	Multiplied by

4. Colors

What is your favorite color?

My favorite color is

Black

Gray

White

Brown

Purple

Pink

Violet

Blue

Green

Yellow

Red

Orange

Light

Dark

Shadow

Tone

Rainbow

Colorful

5. Days of the week

How many days are there in a week?

What day is it today?

Today is monday

What day was it yesterday?

Yesterday was sunday

Is it monday or tuesday now?

What are you doing on the weekend?

On the weekend I...

Today

Tomorrow

Yesterday

The day before yesterday

The day after tomorrow

Week

Weekend

Monday

Tuesday

Wednesday

Thursday

Friday

Saturday

Sunday

6. Months

What month is it now?

It's January now

What was the last month?

Last month was December

What is the next month?

Next month is February

What season is it now?

Now it's summer

Current

Last

Next

Previous

January

February

March

April

May

June

July

August

September

October

November

December

Seasons

Winter

Spring

Summer

Autumn

7. Time

What time is it?

It's one o'clock

It's twelve

It's quarter past three

It's quarter to nine

It's half past six

When were you there?

Two hours ago

When will you be here?

In two hours

To be late

I'm sorry I'm late

Time

Schedule

Second

Minute

Hour

Quarter

Ago

Within/in

Morning

This morning

Day

Today

Noon

Afternoon

This afternoon

Evening

This evening

Night

Tonight

8. Habits

When do you wake up usually?

When do you go to sleep?

I go to sleep at eleven o'clock

Are you busy?

I am very busy

To brush one's teeth

To fall asleep

To lie down

To shave oneself

To shower

To sleep

To snore

To wake up

To wash oneself

Busy

Early

Always

Daily

Monthly

Normally

Usually

Constantly

Sometimes

Routine

Late

Never

Weekly

Annually

Regularly

Rarely

Previously

9. Weather

How is the weather now?

It's hot (outside)

It's cold (outside)

I'm hot

I'm cold

What's the temperature now?

It's 30 degrees now

To rain

To snow

To melt

To flood

To shine

Rain

Snow

Slush

Flood

Sun

Forecast

Temperature

Warm

Dry

Clear

Windy

Bright

Sky

Lightning

Hurricane

Umbrella

Degree

Cold

Humid

Cloudy

Slippery

Dark

Storm

Tornado

Earthquake

10. Countries

Where are you from?

I am from

Where do you live?

I live in Helsinki

I'm moving to USA

What is the capital city of ...?

How long are you staying here?

I'm staying here for five days

To live

To move

To visit

To invite

City

Capital city

Here

There

Stereotype

Continent

Country

North America

USA

South America

Brazil

Europe

United Kingdom

Asia

China

Africa

Egypt

Oceania

Australia

Antarctica

India

Antarctis

Italy

11. Languages

Do you speak ...?

What is your mother tongue?

My mother tongue is ...

What languages do you speak?

I speak a little bit of ...

Can you repeat, please?

How do you spell it?

Can you understand me?

I don't understand you

Can you speak more slowly?

How do you say ... in ...?

Useful

Foreign

To learn

To begin

To repeat

To improve

To make mistakes

To remember

English

French

German

Spanish

Italian

Skills

Dictionary

To translate

To spell

To mean

To correct

To get along

To forget

Well

Badly

Fluently

A bit

Chinese

12. Nature

When is the sunrise?

When is the sunset?

To flow

Mountain	Mountain range
Hill	Cliff
River	Lake
Canal	Dam
Pond	Ocean
Sea	Wave
Beach	Island
Land	Coast
Bay	Peninsula
Forest	Rainforest
Tree	Palm
Plant	Cactus
Flower	Leaf
Ground	Soil
Grass	Bush
Swamp	Moor
Desert	Dune
Arctic	Glacier
Valley	Cave

13. Animals

Do you have a pet?

To feed

Pet

Zoo

Endangered

Extinct

Dangerous

Wild

Species

Bear

Bird

Camel

Cat

Chicken

Cow

Duck

Donkey

Dog

Elephant

Fish

Giraffe

Goat

Horse

Hippo

Lion

Monkey

Mouse

Pig

Rabbit

Rat

Shark

Sheep

Snake

Squirrel

Tiger

Turtle

Zebra

14. Phone Calls

Hello (on phone)

Who's there?

It's ... on the phone

Where are you?

Can I talk to ...?

Do you want to leave a message?

To call

To ask

To reply

To talk

To reach

To connect

To press

To wait

To thank

To call back

To answer

To listen

To chat

To contact

To check

To dial

To send

To postpone

One moment

Now

Available

By phone

Cellphone

Meeting

Currently

Later

Not available

By email

Smart phone

Conversation

15. Family

Who is that?

This is my mother

Who are those?

These are my parents

How many people are there in your family?

There are ... people in my family

To raise children

To adopt

This

That

These

Those

Family members

Parents

Father

Mother

Grandparents

Grandfather

Grandmother

Siblings

Brother

Sister

Older

Younger

Aunt

Uncle

Niece

Nephew

Cousin

Relative

16. Relationships

Are you married?

Yes, I'm married

No, I'm not married

Will you marry me?

Yes, I do

To love

Love

To get engaged

Engagement

To get married

Wedding

To get divorced

Divorce

To cheat on

Cheating

Relationship status

Single

Taken

Engaged

Married

Divorced

Widow

It's complicated

In a relationship

Couple

Wife

Husband

Bride

Bride groom

Girlfriend

Boyfriend

Lover

Gift

Honeymoon

Anniversary

17. Characteristics

What is he/she like?

To describe

Friendly	Unfriendly
Honest	Dishonest
Funny	Boring
Good-looking	Ugly
Beautiful	Handsome
Brave	Coward
Crazy	Strange
Dramatic	Emotional
Confident	Desperate
Shy	Social
Loyal	Unreliable
Normal	Curious
Strict	Relaxed
Nice	Horrible
Polite	Rude
Lucky	Unfortunate
Intelligent	Stupid
Interesting	Boring
Hardworking	Lazy
Cruel	Mean

18. Feelings

How do you feel?

I feel

To worry

To stress

To feel

To express

To cry

To scream

Emotions

Emotional

Motivation

Inspiration

Sad

Happy

Angry

Calm

Energetic

Exhausted

Proud

Ashamed

Bored

Excited

Disappointed

Disgusted

Jealous

Anxious

Scared

Worried

Tired

Relaxed

Satisfied

Thankful

Lonely

Confused

Annoyed

Irritated

Frustrated

19. Professions

What do you want to become?

What do you do for living?

What is your profession?

I'm a ...

I work in the office

How much do you earn?

To become

To work

To interview

To earn

To succeed

Profession

Expert

Actor

Athlete

Artist

Barber

Business man

Cashier

Cook

Dancer

Dentist

Doctor

Engineer

Career

Work

An interview

Salary

Successful

Unemployed

Colleague

Fireman

Journalist

Lawyer

Musician

Nurse

Photographer

Pilot

Police officer

Priest

Teacher

Waiter

20. Buildings and places

What is that building?

Where is ...?

Which floor are you going to?

I'm going to the third floor

To be situated

Monument

Elevator

Architecture

Tourist attraction

Floor

Bank

Bus station

Church

Factory

Gas station

Gym

Hair salon

Hospital

Pharmacy

Hotel

Library

Mall

Museum

Park

Police station

Post office

Railway station

Restaurant

School

Shop

Skyscraper

Stamp

Stadium

Square

Supermarket

Theater

Zoo

21. Directions

I'm lost

I don't know where I am

How do I get to ...?

Is it far away ?

No, it's near

Cross the street

Go forward

Turn left

Turn right

Left

On the left

Right

On the right

Continue straight

To turn

To continue

Forward

Further

Ahead

Away

Close

Far

Over

Across

Along

Opposite

Through, Via

Bottom

Side

North

Northeast

East

Southeast

South

Southwest

West

Northwest

22. Position & place

Where is the box?

The box is on the table

The cat is under the box

Put the cat in the box

Take the cat out of the box

To put

To take

To move

To remove

To place

To lift

Up

Down

On

Onto

In

Into

Above

Below

Under

Off

Out of

Toward

Past

Around

Next to

Among

Between

In the middle

In front of

Behind

23. Life periods

When were you born?

I was born in the year …

How was your childhood?

When I was a child ...

I used to ...

To give birth

To be born

To grow

To retire

To graduate

Birth

Child

Puberty

Adult

Senior citizen

Childhood

Generation

Elementary school

High school

Baby

Teenager

Youth

Middle-aged

Retired

Home town

Gender

Middle school

University

24. School and education

To educate

To explain

To teach

To study

To pass

To fail

The education system

Teacher

Principal

Student

Degree

Intern

Internship

Diploma

Thesis

Classroom

Class

Lesson

Test

Essay

Chalkboard

Chalk

Eraser

Calculator

Pen

Pencil

Homework

Notebook

School subjects

Grades

Biology

Chemistry

Computer science

Foreign languages

Geography

History

Mathematics

Philosophy

Physics

Religion

25. Dining

Are you hungry?

I'm hungry

Are you thirsty?

I'm thirsty

How does it taste?

It's delicious

I'm full

To set the table

To wash the dishes

To eat

To drink

To cook

To prepare

To taste

To lack

To mix/stir

To shake

To grill

To burn

To fry

To rinse

To peel

To cut

Meal

Breakfast

Lunch

Dinner

Snack

Dessert

Cutlery

Spoon

Fork

Knife

Plate

A bottle

A mug

A cup

Frying pan

Pot

Can opener

Bottle opener

Corkscrew

Napkin

26. Food

The taste/flavour

Salty

Sweet

Bitter

Gross

Spicy

Mild

Spices

Cooking oil

Sugar

Salt

Butter

Jam

Ingredient

Meat

Steak

Pork

Beef

Mutton

Sausage

Chicken

Eggs

Mushroom

Wheat

Bread

Slice

Pasta

Rice

Salad

Cake

Beverages

Tea

Wine

Beer

Water

Milk

Juice

Coffee

27. Fruits and Vegetables

I'm allergic to

To prefer

Fruit

Apple

Avocado

Banana

Blueberry

Cherry

Grapes

Lemon

Mango

Orange

Pear

Pineapple

Plum

Strawberry

Watermelon

Vegetable

Beans

Cabbage

Carrot

Cucumber

Lettuce

Olive

Onion

Peas

Pepper

Potato

Pumpkin

Spinach

Tomato

28. In the restaurant

What would you like to eat?

I would like to have ...

What would you recommend?

Do you have a menu, please?

Can we have the bill, please?

Together or separately ?

Here or take-away?

To take

To order

To bring

To serve

To suggest

To recommend

To offer

To wait

To enjoy

To include

Waiter

Client

Menu

Medium rare

Well done

Rare

Appetizer

Main dish

Dessert

Speciality

The check

Change

Tip

Included

29. Money

What is the currency in ...?

Where is the ATM?

Can I pay with a credit card?

I want to exchange money

I want to withdraw money

To exchange

To deposit

To save

To lend

To spend

To withdraw

To transfer

To borrow

To pay

Bank

Cash

Paper note

ATM

Currency

Exchange rate

Loan

Tax

Balance

Coin

Credit card

Wallet

Currency exchange

Receipt

Savings

Debt

30. Shops and shopping

I'm looking for

How much does it cost?

It costs 10 euros

What's the price of this?

It's too expensive

To shop

To consume

To negotiate

To bargain

Market

Super market

Cafe

Clothing store

Bakery

Book store

Mall

Bargain

Discount

Sale

Cashier

Shopping cart

Entrance

Opening hours

Open

Closed

Best before

Rotten

The best before date

31. Transportation & Traffic

Where is the bus station?

When does the bus leave for ...?

How many stops before?

Where is platform number 1?

Where can we buy tickets?

Is this seat taken?

How much is a ticket to ...?

The flight arrives at 10

The plane departs at 12

To rent a car

To order a taxi

To take the bus

To park

To commute

To cost

To land

Seat

Station

Public transport

Bus

Taxi

Airport

Ticket

Arrival

To fine

To hurry

To declare

To take off

Taken/vacant

Journey

Train

Tram

Subway

Customs

Price

Departure

32. City environment

Traffic	Traffic jam
Traffic accident	Traffic lights
Parking lot	Zebra crossing
Street	Alley
Downtown	Urban
Suburban	Neighborhood
Slum	Countryside
Highway	Lane
Curve	Corner
Tunnel	Bridge
Plaza	Harbor

33. Vehicles

Do you have a driver's license?

I go by plane/car/ship/bicycle/Motorcycle/bus

I fly, drive, sail, ride

I go on horse/foot

I ride/walk

To change the tire

To fill the tank

To drive

To fly

To float

To accelerate

To overtake

To steer

To sail

To ride

To brake

To repair

Gas

Car

Truck

Scooter

Ship

Plane

Driver's license

Broken

Full

Used

Engine

Van

Motorbike

Boat

Sail boat

Bicycle

Helmet

Car breakdown

Empty

Own

34. Clothes

What is your size?

Where is the fitting room?

My size is..

I will take this one

It fits you

To wear

To fit

To choose

Fashionable

Trendy

Small size

Medium size

Large size

Extra large size

Shirt

T-shirt

Pants

Trousers

Socks

Gloves

Scarf

Beanie

Hat

Cap

Shoes

Sandals

Dress

Skirt

Underwear

Bra

Bathing suit

Bikini

Belt

Tie

Jacket

Suit

Pocket

Hole

Button

Zipper

35. Body

How tall are you?

How much do you weigh?

To sense

To hear

To breathe

To smell

To taste

To sweat

Gender

Male

Female

Head

Face

Eyes

Nose

Mouth

Lips

Beard

Mustache

Ears

Hair

Curly

Straight

Neck

Shoulders

Hands

Arms

Legs

Thighs

Fingers

Toes

Nails

Skin

Abdomen

Stomach

Chest

Breasts

Back

Waist

Butt

Hips

Heart

Lungs

Muscle

Bone

36. Health

Call an ambulance

Where is the hospital?

I have fever

I don't feel well

It hurts here

To inject

To cough

To break

Emergency

Prescription

Pain

Stomach ache

Broken bone

Runny nose

Diarrhea

Dizziness

Menstruation

Migraine

Injury

Sick

Scar

Mental health

Deaf

Pills

To vaccinate

To help

To hurt

Medicine

Vaccination

Therapy

Back ache

Infection

Cough

Vomit

Sunstroke

Pregnant

Headache

Blood

Healthy

Bandage

Well-being

Blind

37. Beauty and hygiene

I need e a haircut

To put on make up

To have a haircut

To shower

Makeup

Nail polish

Ear rings

Necklace

Shampoo

Razor

Toilet paper

Tooth brush

Dental floss

Clean

Deodorant

Scissors

Hair spray

Hair dryer

Lipstick

Jewellery

Bracelet

Mirror

Soap

Shaving cream

Tissue

Tooth paste

Comb

Dirty

Sun screen

Nail clippers

Haircut

Hand lotion

38. House

Where do you live?

I live in a

How big is your house?

My house is 100 square meters big

How much rent do you pay?

I pay rent ... a month

Come visit me

To find

To build

To own

To rent

Modern

Traditional

Tenant

Neighbor

Apartment

Home owner

Castle

Palace

Mansion

Lodge

Single family home

Semi-detached house

Townhouse

Farmhouse

Dormitory

Studio

Cottage

Hut

39. Home

Where is your room?

Come in | To enter

To open | To close

Inside | Outside

Downstairs | Upstairs

Open | Closed

Main entrance | Hallway

Door | Key

Room | Ceiling

Roof | Chimney

Floor | Walls

Bathroom | Toilet

Kitchen | Dining Room

Living room | Bedroom

Yard | Garage

Basement | Balcony

Mail box | Window

Garden | Fence

40. Furniture and interior

To flush the toilet

To clean

To sit

To stand

Air conditioner

Central heating

Heater

Comfortable

Tools

Lamp

Light bulb

Armchair

Sofa

Table

Chair

Bed

Mattress

Shelf

Wardrobe

Clock

Television

Cupboard

Bookcase

Microwave

Stove

Refrigerator

Freezer

Dishwasher

Washing mashine

Toaster

Coffee maker

Vacuum cleaner

Clothes iron

Carpet

Laundry

Washbasin

Tap

Toilet seat

Mirror

41. Materials

What is this material?

What is this made of?

Smooth

Rough

Pure

Proper

Texture

Textiles

Fabric

Leather

Cotton

Silk

Wool

Metal

Steel

Gold

Silver

Diamond

Copper

Ceramic

Concrete

Clay

Dust

Rubber

Wood

Glass

Plastic

42. Traveling

To book a flight

To miss a flight

To travel

To arrive

To return

To hike

To pack

To plan

To depart

To delay

To hitchhike

To unpack

Travel

Tourist

Adventure

Budget

Border

Backpack

Suitcase

Plan

Itinerary

Vacation

Traveller

Local

Abroad

Luxury

Destination

Bag

Luggage

Route

Activities

Danger

Passport

Visa

One way ticket

Connecting flight

Boarding pass

ID

Insurance

Round trip

Layover

Stop-over

43. Accommodation

Do you have vacant rooms?

Do you have a reservation?

When is the check out?

To reserve	To stay
To accept	To book
To check in	To check out
To cancel	To complain

Registration	Reservation
Reception	Lobby
Deposit	Staff
Owner	Guest
Complaints	Noisy
Hotel	Hostel
Camping ground	Brochure

Single bed	Double bed
Extra bed	Blanket
Towels	Pillow
Sheets	Room service

44. Freetime and hobbies

Do you have hobbies?

What do you do in freetime?

My hobby is ...

To paint

To collect

To interest

To turn on

To play

To spend time

To draw

To watch

To turn off

Cinema

Television

Video games

TV program

Play cards

Board games

Sports

Walking

Hunting

Hiking

Jogging

Fishing

Arts

Music

Festival

Reading

Photography

Painting

Concert

Fashion

Writing

45. Sports

My favorite sport is ..

To win

To lose

To practice

To compare

To walk

To run

To swim

To dance

To kick

To pass

Competition

Competitive

Olympics

Crowd

Race

Training

Athletics

Gymnastics

Gym

Bodybuilding

Cycling

Swimming

Basketball

Volleyball

Soccer

American Football

Martial arts

Motor sports

Wrestling

Boxing

Yoga

Dancing

Golf

Riding

Skiing

Scuba diving

Badminton

Tennis

46. Movies

Let's go to the movies

What kind of movies do you like?

I like .. movies

How long does the film last?

Did you like the film?

It's a good film

To last

To rate

Role

Plot

Dubbing

Subtitles

Ticket office

Popular

Movie theater

Movie star

Popcorn

Lemonade

Candy

Genre

Adventure

Action

Animation

Biographical

Comedy

Documentary

Drama

Fantasy

Horror

Martial arts

Romance

Romantic comedy

Science fiction

Superhero

Thriller

47. Computer

What's the password?

To send email

Press the button

Move the cursor

To type

To save

To move

To delete

To press

To backup

To click

To charge

Laptop

Charger

Screen

Mouse

External hard drive

Desktop

Battery

Keyboard

USB stick

Web site

Software

File

Cursor

@sign

Password

Search engine

Document

Folder

Email

Dot

User name

48. Internet and social media

Do you have wifi?

Do you have an internet connection?

The signal is weak

The internet doesn't work

How many subscribers do you have?

Leave a comment

Add as a friend

Add me on FB

Can I add you on FB?

To give a like

To surf

To connect

To publish

To share

To subscribe

To follow

To add

To block

To update

To download

To log in

To register

To tag

To report

To work/function

Internet access

Wireless

Account

Profile picture

User

Username

Home page

Link

Social network

Blog

Hacking

Selfie

49. Photography

Can I take a picture of you?

Can you take a picture of us?

Press this button to shoot

To take photographs

To focus

To frame

To edit

Photography

Videography

Camera

Drone

Lense

Exposure

Underexposed

Portrait

Black and white

Sharp

Bright

Photograph

Video

Digital camera

Equipment

Tripod

Flash

Overexposed

Landscape

Background

Blurry

Dark

50. Music

What kind of music do you listen to?

Can you sing?

Do you play any instrument?

To sing

To listen

To play

To perform

Music genre

Instruments

Musician

Microphone

Piano

Guitar

Drums

Flute

Violin

A CD

Headphones

Radio

Album

Band

Singer

Ochestra

Crowd

Concert

Tour

Song

Voice

Stage

Event

Silent

Noisy

51. Literature

Where is the library?

My favorite author is

I like to read fantasy books

To lend

To return

To write

To read

Author

Cover

Page

Character

Word

Sentence

Article

Biography

Comics

Drama

Epic

Essay

Fairy tale

Fantasy

Folklore

Science fiction

Letter

Memoir

Mystery

Newspaper

Non-fiction

Novel

Poetry

Prose

52. Nightlife

Excuse me, do you have light?

Do you come here often?

To use drugs

To smoke cigarettes

To play cards

To party

To bet

To joke

To warn

To have fun

To get drunk

To lie

To beat

Night club

Disco

Alcohol

Sober

Drugs

Light

Dealer

Pub

Party

Hangover

Drunk

Cigarettes

Ashtray

Stranger

53. Dating life

Would you like to go out with me?

To kiss

To go out

To have sex

To ask someone out

To fall in love

To ignore

To hug

To date

To touch

To hook up

To trust

To hate

Date

Contraception

Birth control pills

Online dating

Relationship

Temptation

Virgin

Surprise

Picnic

Condom

STD

Dating apps

Love affair

Blind date

Sexuality

54. Law

To sue

To suspect

To kill

To steal

To kidnap

To prohibit

To accuse

To rape

To rob

To shoot a gun

Court

Crime

Criminal

Proof

Jail

Innocent

Illegal

Forbidden

Responsible

Violence

Abuse

Killer

Murder

Smuggling

Terrorism

Shoplifting

Pick pocketing

Trial

Criminality

Lawyer

Witness

Punishment

Guilty

Legal

Allowed

Irresponsible

Burglary

Rape

Thief

Murderer

Smuggler

Terrorist

Shoplifter

Drunk driving

55. Politics

To occupy
To govern
To vote
To sign
To belong to
To run for

Government
Agreement
Parliament
MP
Politician
President
Embassy
Ambassador
Ministry
Minister
Bureaucracy
Party
Referendum
Elections
Representative
Voter
Democracy
Capitalism
Monarchy
Dictatorship
Socialism
Communism
Campaign
Nomination
Citizenship
Citizen
Political
Human rights
Federal
Royal
Administrative
Autonomous
Minority
Majority

56. Economy and finance

To risk

To trade

To import

To export

To invest

To profit

International

Company

Monopoly

Market

Stock market

Commodities

Stock

Bond

Natural resources

Oil

Recession

Policy

Employment rate

Jobless

Sector

Commercial

Tourism

Industry

Revenue

Yield

Interest rate

Expense

Entrepreneur

Manufacturer

Contract

Meeting

Sales

Marketing

57. Religion and beliefs

To exist

To believe

To die

To bury

To create

To mourn

To pray

To worship

Christianity

Judaism

Buddhism

Hinduism

Islam

Animism

God

Heaven

Devil

Hell

Angel

Miracle

Saint

Prayer

Sin

Death

Soul

Faith

Spiritual

Religion

Magic

Religious

Grave

Tombstone

Burial

Rebirth

Missionary

Follower

Holy book

58. Science

To invent

To research

To measure

To weigh

To determine

To indicate

To experiment

To develop

To bend

To dig

To evaluate

To analyze

To classify

Scientist

Genious

Inventor

Researcher

Idea

Invention

Atom

Nuclear

Cell

Molecule

Mass

Matter

Electricity

Invisible

Technology

Robot

Data

Fact

Genetics

Laboratory

Hypothesis

Theory

Magnetic

Field

Fluid

Liquid

Solid

Gas

Particle

Properties

Evidence

Research

59. Space

To discover

To explore

To observe

To orbit

Universe

Galaxy

Big bang

Explosion

Star

Constellation

Milky Way

Solar system

Asteroid

Meteorite

Black hole

Gravity

Astronomy

Light year

Orbit

Satellite

Telescope

Horoscope

Discovery

Observatory

Astronaut

Spacecraft

Explorer

Planet

Moon

Mercury

Venus

Earth

Mars

Jupiter

Saturnus

Uranus

Neptune

Pluto

60. Global warming

To eliminate

To stop

To reduce

To recycle

To affect

To recognize

To use

To produce

To deteriorate

Climate change

Global warming

World

Environment

Atmosphere

Greenhouse gas

Carbon neutral

Carbon friendly

Carbon footprint

Diversity

Destruction

Pollution

Damage

Energy

Solar energy

Solar panels

Waste

Harmful

Harmless

Under threat

Irreversible

Carbon dioxide emissions

Renewable energy

Sustainable development

61. History

To colonize

To conquer

To inhabit

To unify

To destroy

To rule

To attack

To defend

To emigrate

To immigrate

To declare

To become independent

Historical

Ancient

Flag

Independence

Empire

Kingdom

Republic

Dictatorship

Invasion

Colony

Pyramid

Ruins

Army

Military

War

Revolution

Civil war

World war

Era

Century

Immigration

Emigration

Immigrant

Emigrant

Refugee

Slavery

Plague

Crisis

62. Pronouns

I	Me
You	You
He	Him
She	Her
It	It
We	Us
You	You
They	Them

My, mine	Myself
Your, yours	Yourself
His	Himself
Her	Herself
Its	Itself
Our, ours	Ourselves
Your, yours	Yourselves
Their, theirs	Themselves

Each other	One another

63. Pronouns 2

What

Which

Where

When

Why

How

Who

Whom

Whose

Whatever

Whichever

Wherever

Whenever

Whoever

None

Every

Nobody

Everyone

Nowhere

Everywhere

Never

Always

Nothing

Everything

Neither

Each

Any

Some

Anyone

Someone

Anywhere

Somewhere

Anything

Something

64. Top 30 Verbs

Can
Shall
Must
Might
Would

May
Will
Could
Should
To be

To come
To do
To have
To want
To know

To go
To make
To need
To like
To understand

To say
To see
To get
To tell
To happen

To speak
To look
To give
To think
To put

65. Important Verbs

To try

To show

To keep

To let

To pull

To expect

To seem

To hold

To lead

To provide

To consider

To appear

To remain

To fall

To organize

To decide

To allow

To require

To finish

To matter

To ask for

To deserve

To take back

To loose

To carry

To hurt

To demand

To concern

To obtain

To propose

66. Comparative

Good	Better	Best
Bad	Worse	Worst
Far	Further	Furthest
Close	Closer	Closest
Old	Older	Oldest
Young	Younger	Youngest
New	Newer	Newest
Beautiful	More beautiful	Most beautiful
Little	Smaller	Smallest
Big	Bigger	Biggest
Much	More	Most
A little	Less	Least

67. Comparing

Good	Bad
Ugly	Beautiful
Strong	Weak
Big	Small
New	Old
Young	Old
Tall	Short
Long	Short
High	Low
Fat	Thin
Light	Heavy
Soft	Hard
Easy	Hard
Full	Empty
Hot	Cold
Cheap	Expensive
Alive	Dead
Safe	Dangerous
Poor	Rich
Quiet	Loud
Right	Wrong
True	False
Clean	Dirty
Positive	Negative

68. Imperative

Read

Write

Stop

Help!

Be nice

Tell me

Visit us

Stand up

Sit down

Come here

Go away

Run, hide

Call me

Get out

Don't disturb

Clean your room

Please be quiet

Wait for me

Have a nice trip

Pass the salt, please

Push/pull the door

Close/open the window

69. Conjunctions 1

After ....

Again ....

All ....

Almost ....

Also ....

And ....

Because ....

Before ....

Both ....

But ....

Enough ....

Few ....

Finally ....

Later ....

Little ....

Many ....

Maybe ....

Much ....

Not yet ....

Or ....

Other ....

Really ....

Same ....

Since ....

So ....

Soon ....

Still ....

Then ....

Too ....

Very ....

Yet ....

70. Conjunctions 2

Actually

Afterwards

Although

Another

As a result

As soon as

Besides

Despite

Either

Even though

For example

From here on

Furthermore

However

In fact

In the meanwhile

In case

Indeed

Instead

Neither

Nevertheless

Often

On the contrary

Therefore

Thus

Unless

Within

As ... as

Both ... and

Either ... or

If ... then

Neither ... nor

Not only ... but also

Whether ... or

71. Adverbs

Absolutely

Apparently

Automatically

Basically

Carefully

Certainly

Clearly

Completely

Deliberately

Definitely

Especially

Evidently

Exactly

Extremely

Generally

Honestly

Hopefully

Immediately

Literally

Mostly

Nearly

Obviously

Originally

Particularly

Personally

Physically

Possibly

Probably

Relatively

Seriously

Suddenly

Slightly

Totally

72. Structures

According to

By the way

Of course

Referring to

I agree

I am sure

It depends

I feel like

I have to

I think

It seems

In vain

In addition to

Unfortunately

In terms of

In my opinion

I am going to

I can

I disagree

I guess

I like

I want

It is true

Like this/that

Never mind

Are you sure?

I would say that

I don't think so

You are right / wrong

Once upon a time

From my point of view

If I understand correctly

It would be fun to

CPSIA information can be obtained
at www.ICGtesting.com
Printed in the USA
BVHW011420170520
579806BV00009B/513